ON WRITING THE BIOGRAPHY OF A MODEST MAN

LONDON : HUMPHREY MILFORD
OXFORD UNIVERSITY PRESS

LEBARON RUSSELL BRIGGS AT SEVENTY-THREE

ON WRITING THE BIOGRAPHY OF A MODEST MAN

By

Rollo Walter Brown

Cambridge · Massachusetts
Harvard University Press
1935

FIRST EDITION, MARCH 1935

PRINTED IN THE UNITED STATES OF AMERICA
BY THE HARVARD UNIVERSITY PRESS

AN ADDRESS DELIVERED BEFORE THE RADCLIFFE CLUB OF NEW YORK CITY ON NOVEMBER 25, 1934, AND BEFORE THE HARVARD CLUB OF NEW YORK CITY ON DECEMBER 11, 1934, AT MEETINGS HELD IN HAPPY REMEMBRANCE OF DEAN BRIGGS

ON WRITING THE BIOGRAPHY OF A MODEST MAN

DEAN LEBARON RUSSELL BRIGGS BECAME A person of interest to me the moment I first saw him. I had straggled into Harvard from a small college in the Middle West. While I sat outside the Dean's office waiting my turn to see what he was like and to ask permission to take his course in writing, one of the students ahead of me left the door slightly ajar when he entered. Through the wide crack I could see directly to the desk in the opposite corner of the office. A tallish man of forty-five or fifty stood beside the student and looked with cheerful, intent face at a long list of some kind which he was holding very awkwardly in one hand. His neutral-colored hair was much kicked up, he had cut himself very noticeably in front of his ear when he had shaved that morning, and his eye-glasses were on the point of toppling head foremost over the end of his nose.

When at last I had opportunity to go in and tell him why I was there, he listened patiently, though with a show of nervous tingling, and then said with great kindliness and faltering, as if his words caused him pain, "I hate to tell you, but I cannot possibly take another man. I ought not to have taken the last one, but he seemed to think I had half-promised him a place; so I added his name. But I must not add another. I'm afraid I'll never be able to read all the themes, as it is."

He saw how great my disappointment was, and as I turned away he offered a consoling word: "If you should be here to-morrow morning quite early and anybody has dropped out, perhaps I might give you the place—though I ought not to give it to anyone."

I was at University Hall an hour before the offices were supposed to be open. But the outer doors were unlocked, and I slipped in, climbed the granite stairs, and took my place in the chair nearest the Dean's door, so that nobody could possibly get ahead of me. After I had sat there in the quiet for fifteen or twenty minutes, I was startled by what seemed to be a movement in the office. I listened, got up from my chair, listened again, heard nothing, and then, in order to make sure, knocked with relaxed knuckles.

"Come!" the cheerful voice of the day before called.

I pushed the door open hesitantly. He stumbled up from his chair, hooking his glasses off on his thumb as he did so, and met me half-way across the room. His smile as he spoke was very friendly, but touched with shrewdness: "You did come early, didn't you?" Then he seemed to think it necessary to explain his own early presence. "I like to get down here and work awhile at my desk before the rush comes."

He picked up his class list and scanned it as though he were really not seeing it. "But, are you quite sure you want the course?"

I was sure enough. If he had told me that he had decided to offer the course in Vedic Sanskrit, I could have found reasons for thinking Sanskrit necessary to an interpretation of twentieth-century America. "Oh, yes, I want the course."

He looked at the list again. "Very well, then. A man telephoned late yesterday evening that I need not count on him."

He first revealed an interest in what I was trying to do when I one day wrote a short theme in which I bewailed the fact that the preliminary requirements for graduate study in modern English at Harvard were so many that by the time a man had met them all he could hope to have only just about time enough left before he died to read Tennyson's "Crossing the Bar." He revealed a more serious interest in me — or perhaps it was concern for me — when I went to him, before the college year was yet old, with a very acute personal problem. He helped me to solve it. And when I thanked him, he thanked me for feeling that he was a person in whom I could confide.

I did not know at that time how many other men he had befriended, or in what ways. I only knew that there was nothing of the ordinary college-office first-aid attitude in his friendship, and I believe he saw that my devotion to him was not wholly gratitude for his having helped me out of a tight place. At any rate, he quite casually encouraged me to spend time with him when I could not understand how he could have free time for anybody. Sometimes he invited me to walk homeward with him at the end of the day, and induced me to talk while he swung along in weary content with his green book-bag over his shoulder and his crushed brown hat pushed back ever so slightly. He invited me to drink Matzoon with him—a pungent, substantial soda-fountain beverage that to me always tasted like Walt Whitman's poetry. He hustled me over to Jimmie's with him for hasty "perpendicular" lunches. He told me stories about President Eliot, for whom he had an admiration that would have been reverence had he not possessed quite such a ranging eye for comedy. He told me stories about other Boston celebrities, about Plymouth farmers, about students — mentioned anonymously — who had written this or that astounding specimen of English. He made incidental mention of John Donne, Lovelace, George Herbert, Crashaw, Herrick, Waller, and many another poet with whom I had little or no acquaintance. He confessed to me that while he was not able to share my enthusiasm for the whole of Walt Whitman, he had never been able to finish reading "When Lilacs Last in the Dooryard Bloomed" and especially "O Captain! My Captain!" with dry eyes. He made sly, humorous comments on athletics, on coeducation, on practices in university administration, and was certain that the country would greatly profit if all educators in high place should spend more time reading *Mr. Dooley*.

Through fifteen years immediately after I left Cambridge, he remained not only my greatest friend, but my surest proof that not all of life is futile. I heard from him rather frequently, and at least every other summer I made my way back to Cambridge at Commencement time to see him. I tried never to be bothersome, and I do not believe he thought me so. He could not fail to be aware of my chief single reason for being back, and simply absorbed me into whatever he was doing, without letting my presence disturb him. Sometimes when he sat in the vast Faculty Room — usually with one foot tucked under him — signing piles of diplomas and at the same time asking me what I knew about this or that man of my generation, or telling with quiet mirth what the watchful ladies of Cambridge had most recently proposed that he do in the interest of morality or less exhausting athletic sports, I often wondered if he did not sometimes put his name in the wrong place, or sign it twice on the same diploma. But I never heard of instances. Capacity to absorb people or things into the current of whatever he happened to be doing was his one great natural equipment that made up somewhat for his utter lack of the labor-saving instinct. It came about, then, that I was with him on all sorts of occasions, and at all sorts of times when he was engaged in such non-academic enterprises as doctoring sick dogs, carrying flowers to people who needed flowers, buying meat — as if it were a religious rite — proving to grocerymen that they were trying to collect a bill he had already paid, looking over attractive horses hitched along the street to see if they were sound, trying to secure employment for colored servant girls, or packing up to go to the country.

Yet it was not until I was more or less by accident back in Harvard as a teacher that I began to think of writing about him. Before, he had been only an incomparable friend. But when I had full opportunity to see him day after day in his later years as he rushed up and down the stairs of University Hall, counseled with men about all sorts of troublesome matters, fought for decent athletic sports, kept a lingering eye on Radcliffe College, pleaded for open-mindedness, writhed in suffering because of the hatreds and prejudices the World War had left in men, and yet was able to command unlimited resources in cheerfulness and serenity, he came to be for me a kind of beloved phenomenon.

My great fear was that he would modestly refuse to be written about. That matter had to be cleared up first of all. We both had classes on Tuesday and Thursday at two o'clock, and since he was usually ready for relaxation after working from early morning until three in the afternoon, and since our classrooms were not far apart, we often spent a quiet half-hour together, and sometimes continued our discussions on the way home, although he more frequently went back to his office for another turn there. While he lounged against a conference table, or stretched full length upon it—he said he liked the feel of a hard table, or the floor, against his shoulders when he was tired—or hung by his knees and his elbows back over the long straight radiator across the bow window in his room, we discussed everything from world chauvinism to the most personal dreams and disappointments. Yet I never felt that I dared mention anything so hazardous to our intimacy as the proposed book.

All winter I waited. Then late one day

in the early spring after I had passed him as he talked with a colleague in front of the Law School, he overtook me and we crossed Cambridge Common together. He looked worn, and as we walked along he told me about a harrowing problem that had arisen in his office. It had kept him from sleeping, despite the fact that he felt sure of his ground and saw ahead clearly. I am still quite unable to say why the circumstances seemed auspicious, but before we reached the corner of Brattle Street and Ash where he lived, I felt that I could make some direct mention of what I had in mind.

As we lingered on the street corner—we were already discussing matters far from serious—I said to him: "Before I go, I should like to ask you a question. In twenty years, you know, I have asked you some pretty stiff ones. But this time I expect to bowl you over."

He laughed, leaned against the one-way sign at the edge of the sidewalk with his green book-bag across his shoulder as if he were a target, and commanded: "All right. fire away!"

I told him that I wished to write his biography, and that I meant to do so unless the idea were so utterly distasteful to him that he forbade me to proceed.

"But," he protested, "you write books about celebrities, and I am not one."

"You know I studied law once for a couple of years," I countered, "and I feel ready to argue before the court that you are not a competent witness on that point."

He went into detail. He could not understand why people were disposed to magnify what he had done. Harvard men insisted on honoring him for doing his simplest duty. And now that he had retired from Radcliffe, they were doing much the same thing over there. It was nothing short of inappropriate. Suddenly he broke off with a question which, try as he might, he could not quite take seriously himself: "You didn't have anything to do with naming this new Radcliffe building for me, did you?"

Before I knew it, the conversation had got completely out of hand and he was telling me a story about Mrs. Agassiz.

"Won't you come in and have a cup of tea?" he asked as he moved to go toward the house.

"But you are getting away from the subject," I insisted.

He glanced straight at me with a half-fearful smile playing over his face and said: "I am sure you do not expect me to give my approval. But you are a free moral agent, and, of course, you can do as you like."

He moved again to go. "But are you sure you can't come in and have a cup of tea?"

I pleaded the necessity of getting home. I did not want him now to grow panicky over my presence. But before I left him I sought hurriedly to fortify him against dangerous afterthoughts: "You see, I possess such quantities of material already that I shall not need to bother you for much. You mustn't think every time you see me coming in your direction that I am out for copy!"

"No, for I do want to have a few unembarrassed conversations with you yet before I die."

"But I may have to come to you for certain information about your father, since you are the only person who possesses it."

"All right. I'll tell you as much as you wish about him. I can talk about him without any trouble whatever."

II

Two or three weeks later I stopped at the house to see Mrs. Briggs.

"Has Mr. Briggs said anything to you about a project of mine?" I inquired.

"No," she answered. "What is it?"

I told her, and explained that I might have to come to her for photographs and other strictly family materials. When I was ready to go she asked: "Now, shall I speak to him and let him know that you have come to me, or shall I wait to see if he mentions the matter himself?"

I left everything to her wisdom.

Weeks later she told me that one day when she and the Dean were alone together —in the country, I believe—he asked shyly: "Do you know that Mr. Brown is going to write a book about me?"

"Is that so?" she asked.

"Yes," he replied. And there the conversation ended.

In the autumn I went to the house one rainy Sunday to go over the early photographs with Mrs. Briggs. The Dean heard familiar voices and came downstairs.

"Hello, there!" he called as he entered the room, quite unaware of what we were doing. When he saw the pile of photographs in Mrs. Briggs's lap he said in a smiling effort to throw off his confusion: "I see where I've got to take a trip to Europe, if my wife is going into this conspiracy, too."

"Good!" Mrs. Briggs exclaimed quite casually. "I'd like to go to Europe myself—after I've found the photograph I am looking for here."

He retreated uncertainly toward the dining-room and then toward the kitchen.

The next morning I met him in the College Yard. What he wished to know at once was whether I had time to go up to his office and see a letter he had received about a young man whom we both knew.

When I was ready to leave, he came to the doorway, stepped two-thirds of the way out, reached back and hooked one foot round the edge of the door to hold it close against him, and then said very seriously in a lowered voice: "I am sure you know that I am not unmindful of what you are doing — though I do think you are on the wrong track — but I live in constant dread that somebody will imagine I 'inspired' you to write this book, or will suspect that I have collaborated with you."

I told him that he was the only man on the face of the earth who would ever entertain these possibilities.

Some days later when I met him as I crossed the Yard, he asked somewhat fearfully, "You were not coming up to —"

"No, not at all," I assured him. "You must not let yourself believe that just because I am busy with this book I have come to regard you as a field of research."

A sly twinkle displaced the half-fear which his eyes had expressed. "No, I shouldn't care to feel that I was the subject of a Ph.D. thesis."

I sought to assure him further. I should need to bother him, I explained, only two or three times in all. For these few occasions I meant to give him ample forewarning. He could know, then, whenever he saw me, that unless he had received a warning, the book would not be mentioned.

"But you still wish that information about my father, don't you?" he asked, as if I might have forgotten something.

"Oh, yes; very soon."

"All right. Let me know — the sooner the better."

Not long afterward, I gave him warning and then went to see him. "I think we'll not stay in here," he said, glancing about as if there might already be sharp-eyed onlookers in every corner. He led me up narrow stairs to the very top of University Hall, and there in a tiny room where the heavy hewn roof timbers hung close over our heads and the afternoon sun shone in at the dormer windows, we had a long session. But it was not long enough. We had to go back for another the next afternoon. Then for months we walked together, and, when spring came, went to baseball games together, as if no book had ever been contemplated. Once when he came into his office from a class and found me going over some material with his secretary — some quotations from his annual reports that she had wished to copy as her contribution to the volume — he looked nervously toward us but said nothing, went to his desk in the other corner of the large office, and gave himself with a great show of concentration to his work.

For months I was busy with trips to Yale, Princeton, New York, and wherever else his vitality had left biographical material. The college year came to an end and he had retired from active duty before I gave him my second warning. I stopped at the house late on the afternoon of Commencement Day to congratulate him on the generous fund that the alumni at their annual meeting that afternoon had set aside for him and Mrs. Briggs.

"Did you know that that was going to be done?" he asked a little excitedly.

"Yes, I knew."

"Well, I didn't — didn't know a thing about it until to-day."

"You were not supposed to know."

"So it seems!" And quickly he turned to the subject of packing. "We want to get to the country as soon as possible. But, you see, the packing will take longer this time."

"Then I had better come before you get into it. You are hereby warned."

A day or two later when I stopped, he called down from upstairs: "I'm packing some extra books that are up here. I'll be down in about two minutes."

I waited in the hallway. Soon he appeared at the head of the stairs without a coat, with his collar loosened, with his old blue felt slippers creeping from under his heels, and with three or four books clutched in hands held stiffly out in front of him.

"It will require only a half-minute," I explained, "and it will be easy."

He came to within two or three steps of the foot of the stairs with all the books still in his hands.

"Of course," I said in an effort to put him at ease if possible, "I am to be responsible for all opinions and interpretations I express. But I wish to ask you about a matter of fact that no one seems ever to have reduced to a few words—and I wish to do so. I have in mind saying, unless you correct me, that in all your efforts in college discipline you strove, first, to help the student disciplined, and not merely humiliate him—"

"Yes," he answered very simply as he stood on the second step, still holding on to his books.

"And that you sought to administer discipline so that the faculty would be as free as possible to do its proper work—"

"Yes."

"And finally, that you tried to develop a sentiment among students which would render discipline less and less necessary—"

"Yes."

He stood for a moment oblivious of everything about him—a pleasingly disheveled old man with books clutched in hands smudged from packing—and looked away as if he were seeing back to the beginning of his career.

He started up, and then smiled. "Yes, that is about what I tried to do. If you are going to say anything, you'll be safe in saying that."

Then abruptly he asked, "Don't you think this picture of President Eliot"—right at our shoulders in the hallway—"is about the best of them all?" And he bethought himself to sit down and put the books on the step beside him before he entered into a discussion of all the photographs and etchings and lithographs of President Eliot that he remembered.

Even in correspondence he revealed this

watchfulness to get away from the subject when the subject happened to be himself. I once wished additional information about certain of his favorite horses, cats, dogs, dahlias, and railway locomotives, and since one of his greatest amusements was what he called "the new education by questionnaire," I warned him that I was about to send him a questionnaire from New Hampshire. Then I sent him one on "plants, animals, and mechanical devices." Evidently he found it amusing. For he not only filled in the answers, but gave me such incidental reminders as "Pippa was a mare, not a cat," and begged to call my attention to the fact that the splotch of ink on page 1 was part of his "contribution." But from even this revelation of himself he had to turn quickly to the remark that, seriously, he could not believe his pastimes of interest to any reading public, and then to the suggestion that it would be well to come down to Halfway Pond about October tenth, since the Plymouth colors were usually at their best at that time.

When we made this trip to the Pond in October (1925) I warned him for what I assured him was the final time. So many apocryphal stories about him had been set adrift from Harvard Square that I wished to check over episodes to make doubly certain that nothing apocryphal found its way into the book. We sat on the piazza, and while the Dean gazed across the Pond and patted the head of an old dog that insisted on keeping its nose against his knee, he listened as I summarized from the manuscript, and gave his "Yes" in subdued voice.

Once he stopped me. I had told of how two solicitous Cambridge ladies came to him to ask him to prevent a certain cruelty to animals. The University of Florida baseball team was coming to play Harvard, and some student publicity expert gave out the impressive information that Florida's mascot, a drowsy young alligator, would lead the procession from Harvard Square to Soldiers Field. These two ladies thought it would be inhuman to make a poor alligator march that far in the hot sun.

"Where did you get hold of that?" he asked in surprise.

"Oh, I picked it up," I replied. As a matter of fact, one of his sons had put me on the track of it.

"Well, now," he said, squirming about in his chair, "I suppose the things a man would leave out himself are just the things the readers are looking for. But those two ladies are still my neighbors, and it might be safer if you did not print that while I'm alive."

He smiled; then he laughed. "They came

to me with what President Eliot would call the highest of motives; but I can't believe that they had ever made a very close study of an alligator's cadence."

We went on. Once again he stopped me. "Not that that's apocryphal, either," he said. "The man who got himself into the mess would be only too glad if it were, I've no doubt, especially since he is now in politics. But I gave him my word that I would not persecute him, and if you publish the incident, even though you have omitted his name, he will recognize himself and think I have broken my promise."

He did not insist that I leave the incident out. But I was able to find another that served quite as well—in truth, better—and used it instead.

When I had completed the list he seemed greatly relieved. "And now, since you are so ready to have the blood of all the opinion and interpretation on your own hands, I think I'll see if Mrs. Briggs and Mrs. Brown aren't ready to have me hitch up for a drive around the Pond."

III

The Briggses were in Europe at the time the book appeared early the following summer. Mrs. Briggs asked me to send her copy to Halfway Pond, since they expected to go directly there from the boat when they returned in the early autumn.

One morning soon after they arrived, the Dean called me by telephone. He was in Cambridge, and wished to know first of all how my wife, who had been in the hospital, was coming along. He was up for the day only—for the formal opening of the year at Radcliffe College—and had nothing in particular to do for the next forty-five minutes—till quarter to ten. But perhaps I was just ready to go to my work.

I hurried off to see him. I had not expected him to read the book, and had said so in the preface. But now that he had had opportunity to read it if he wished, I was suddenly quite as jumpy as ever he had been over any of my "warnings." I thought of a hundred passages that might not withstand his critical scrutiny.

He was exceptionally cheerful. He wished to know if I had ever visited Taormina. He had found so many glories there that we had no time for anything else before quarter to ten.

He gathered his hood and gown together, crushed them over his arm, shoved his mortar-board down into the angle of his elbow on top of the hood and gown, put on a spreading black hat which, he explained unnecessarily, he had borrowed—because he had only a straw hat down in the country—and with this over-large black hat close down on top of his ears, and his gown dragging on the sidewalk and catching on every rough-edged brick, he directed our way toward Radcliffe. As we walked along, he said calmly, "Mrs. Briggs has read the book, and likes it; and you may have already heard from the children and my sister. I peeped into it, but I did not quite have the nerve to read it."

Whether he ever read it I do not know. I am sure, however, that he had not done so about two years later, for one day at a baseball game he told me, as if he were imparting new information, something that I had discussed at length in the book.

Nor did he ever mention the enterprise to me again, except once. On another afternoon, as we made our way toward the ball park in great leisure because of his slight temporary lameness, he asked me as we walked under the young sycamores if I had ever regretted that I turned from teaching. I told him that I had not; that I had enough writing planned to keep me busy for at least twenty years, and expected to live or die by it. A warm smile lighted up his face, and then, all the while looking straight ahead, he remarked, "And I believe you may do pretty well at it, now that you are through writing biographies of unimportant people." When anyone else spoke of the book in our presence — and it was not unusual to have great admirers of the Dean do so — that part of the conversation was allowed to die as painless a death as possible. When some Harvard graduate sent me a copy of the book and asked not only that I autograph it myself but that I have Dean Briggs do so, I usually wrote an inscription absolving the Dean from all complicity, took the book to him, opened it without comment at the fly-leaf, and then glanced at books in the little downstairs study while he wrote something himself. When he was through and had handed the book back to me without a word, he often made some such remark as "I suppose you never knew him. He was too far back — class of '97, I believe. All three of his boys came to Harvard sooner or later. Their mother was the daughter of a classmate of mine in '75 — and one of the loveliest women I have ever known."

IV

In the eight years of his life after the book was published, I saw more of him than before. He was busy much of the time as a member of committees of the Board of Overseers at Harvard, and he was active in many other groups that concerned themselves with semi-public enterprises. Yet he had greater freedom in using his time than he had ever before enjoyed. He had entered into an agreement with one of the large publishing houses to write his memoirs, but when he came face to face with the actual writing, he asked to be released from the agreement. "I saw," he explained to me one day in rejoicing over the still greater freedom that he hoped he was now to have, "that if I wrote those memoirs and did not conceal a part of the truth, I was going to cause pain to people whom I would not hurt for anything in the world — and get myself into a hornet's nest, in the bargain. So I think I'll just concentrate on baseball for the rest of my life, and let it go at that."

He usually had supplies of tickets from the Boston Major League clubs — he never got through explaining that while they were given to him he had not begged them — and he was always looking for somebody who could go to games with him. He liked to be there ahead of time. "It is never quite a whole game for me, you know," he always explained if he feared I might be late, "unless I can see the attendants make the white lines of the batter's box before play begins." Since my writing-day is over rather early, it came to pass that we went much together — as often as two or three times a week. At seventy-seven he was the best-informed fan I have known. Constantly he set sports correspondents and baseball managers straight on college and league scores and players of ten, twenty, thirty, and forty years ago.

In his inside coat pocket he carried a leather case that served as a depositary for unusual scores. But it served also as a depositary for anecdotes from *Punch*, new poems that he liked, clippings about former students, photographs of his grandchildren, lists of groceries that he was to order or take home with him, ideas that he thought worth jotting down. I never did see him get all the way through the contents of that case, and only on a few occasions did I ever see him find what he was looking for. But he invariably found something else that was interesting, something that started a train of stories, memories, observations. He did not repeat the same story to the same person after the manner of some old men. But he was always afraid that he might be doing so, and was constantly prefacing new stories with, "Did I ever tell you—?" "Did I ever tell you how President Eliot said to me, 'No, don't say he is "low-down"; just say "less sensitive"'?" Or, "Have I ever said to you before—?"

And on the subway train, in Fenway Park, at Soldiers Field, on Larz Anderson Bridge while we rested our elbows upon the coping and watched the crews on the river, his active mind was busy with all sorts of interesting oddments that he had remembered or thought upon. He was so free from guile that it never occurred to him to speak in veiled statements. With wholly untrammeled honesty he discussed men in public life — he remembered the grades they had made in college — he characterized colleagues, he commented upon changing conceptions of culture, he reflected upon new social practices. Sometimes he was caustic, as when he told me of a nationally known editor who bought an article from him and then "lifted" a certain section of it and published it as a part of one of his own signed editorials. "Of course, it was *his* after he bought it!" Sometimes he was distressed by the ethics of men in whom he had had confidence, as when he discussed a Vice-President of the United States who had proved in a magazine article that the Radcliffe undergraduates were a bolshevik lot, by showing that a debating team representing them had one evening argued for the closed shop, and by withholding the equally important information that another team representing them had on the same evening argued for the open shop. "I never can believe he wrote that article himself," the Dean declared. "He must have hired somebody to do it for him and then signed it without knowing what was in it. But let me see, that explanation would not help matters much, would it?" He made an effort to smile, as though he were trying to conceal the hurt of some great, inclusive soreness of body. "I'm afraid the terrors of the closed shop will remain as nothing compared with the terrors of the closed mind." Sometimes he became extremely grave and talked with the nervous, fading voice that troubled him much when he spoke in public. But more often he was swept by irrepressible humor and made his commentaries with invigorating saltiness. Once when the Harvard football team was having an unusually poor season, some players on their way to practice in a strikingly luxurious automobile nearly ran us down as they made the short trip from the north side of the Charles to Soldiers Field. "Perhaps I'm old-fashioned," the Dean observed as we walked on, "but I sometimes wonder"—and his voice became crackling and merry—"if Harvard may ever hope to win games unless we somehow find players who are equal to getting from one end of the Larz Anderson Bridge to the other under their own steam." On another occasion, while we stood by the Charles and surveyed the Graduate School of Business Administration, dazzling in its newness and much white paint, he remarked as a sly smile flitted across his closely checked pink face, "It reminds me of the Spotless Town in the old advertisement for Sapolio."

There were, to be sure, all sorts of occasions when baseball was not responsible for our being together. Often I met him along lower Brattle Street late in the afternoon when he was on his way home with a green book-bag full of provisions, and he half leaned, half hung like a grotesque question mark against the brick front of a grocery store and told me the latest anecdote he had liked — of the British lady, for instance, who had said after heavy storms in the English Channel, "Just think, my dear, for two days the Continent has been completely isolated!" When he was shut in for several weeks of observation before he was taken to the hospital for an operation that his friends feared might not be successful, I sometimes sat with him in the bright room upstairs while he lounged on the wide bed and read to me from the book of charades that he was writing. He seemed to be much more concerned over the way the charades were coming out than over any possible outcome of his period under observation. When finally it was decided that he must go to the hospital, he protested that they must hold off a little until he had finished the last of the charades. "I promised Pottinger that he was to have the manuscript, and he must have it." And after the operation—which turned out to have been unnecessary—just as before, he remained my final court of appeal whenever I was finishing a manuscript, or was reading proofs and had to contend with editors who believed the subjunctive mood ought to be eliminated from the English language. After he had tried a sentence out on himself—he had become eye-minded through a lifetime of theme-reading and had to look at whatever he put to the test — he would re-live a little of his experience as a teacher by digging from the leather case in his pocket, or from his memory, every sort of interesting specimen of idiomatic English, from Dryden's "He was a man stepped into years, and of great prudence" down to the instance he had just found in some recent volume written by a former student.

Yet through all this greater intimacy that grew from what I could not fail to see — and with a certain regret — was a habit of looking on me less as a youth and more as a near-contemporary, I never discovered in him anything that would lead me to revise my earlier interpretation of his life — save in this: the closer I came to him, the more genuine I found him. The high level on which he lived was his natural level. The irrepressible inclination to see the ridiculous in so many things was his natural appreciation of the vast chasm between what men pretend and what they achieve. The glow of good will which shone in his face was the natural expression of his own great humility and his own great beneficence. With Miguel de Unamuno he cried out through everything he did: "Warmth, warmth, more warmth! For men die of cold and not of darkness; it is not the night but the frost that kills."

This fundamental warmth of spirit infused every other quality of his character. No matter how his fellow beings chanced to come in contact with him, they discovered in him very soon something that was friendly to life. Colleagues said, "Yes, he's one of the fellows Briggs helped through college." Professors at Yale said, with the least trace of pleasure in their skepticism,

"But do you think the Harvard of to-day produces men with the sympathies of Dean Briggs?" Harvard students far from home said, "We were feeling low last night and went over and spent an hour with the Dean —and his cat Joshua." Radcliffe graduates said, "We wouldn't for the world have had him a nice, sleek, impervious president in a morning coat." Policemen on the lower Brattle Street beat said, "And don't you suppose we could tell you about one or two ourselves that he kept from going plumb to hell?" The Irish housemaid who had long been with the family said, "And once in the country when he was taking me to ten-thirty Mass, and the door of the old machine came open a dozen times, he slammed it hard and said, 'Confound it!' And said I, 'Oh, Mr. Briggs! Now I've learned what I've been wanting to know all these years: you are no angel!' And then for days after, when he was in the house he'd say softly but so that I'd be sure to hear him, 'Well, confound it!' And I'd say, 'Oh, try to patch things up if you want to, but I don't believe you can ever get forgiven for anything as grievous as that!'" And Harvard men of all ages below fifty or sixty said when they saw him trudging along in front of the baseball stands with his big yellow blanket on his arm, "Why, there's the old Dean!" Then they gathered round him in such numbers that the policeman had to ask them—in the least obtrusive manner he could invent—not to block the way behind the catcher's net, and the Dean was unable to get to his accustomed seat along the third-base line until the first or second man at bat was out.

It was this warmth of spirit which resulted in such vast accumulations of renown and affection that to many he seems more like a legend than a man who took his departure only one morning last spring. "He is not real!" "Men like that do not exist!" "He is too good to be true!" "He is somebody's creation!" And now that I never meet him in Harvard Square or in the College Yard—though in some ineradicable way I am constantly expecting to do so—I sometimes wonder myself: Is it possible that I once walked in the full light of day in this matter-of-fact, turbulent world with such a man as that?

There was about his life something beyond belief. But it did not result from any unreality. Rather it resulted from a greater reality than most men are prepared to see. He struggled. In his childhood he had had the uncommon experience of living without respite among people who were beyond him in size and years. His brother was older; his two half-sisters were older; and because of his aptitudes he was in school classes and school games where everybody was older. Always he had to be exercising his mind earnestly—painfully, he sometimes confessed—in an effort to come up to something that was a little out of reach. This unquiet, exploring mind stayed with him. With the desperation which the hungry reveal he sought to draw near to all sorts of remote worlds, all sorts of new understandings, all sorts of subtle intimations, which to the less persistent must forever remain far off and nebulous. In consequence, he seems unreal because he lived so close to truth.